EVOLUTION

FACTS versus FICTION

Gene Keith
October 2014 – Revised 2019

Why should Christians concern themselves with the theory of evolution or intelligent design?

After all, if we believe God created everything, what difference does it make if He created the world instantly or used the process of evolution to develop everything gradually over long ages?

Either the Bible is true when it says a personal God created the heavens and earth and all it contains, or evolution is true. No one can have it both ways."

Table of Contents

Introduction

Why should Christians concern themselves with the theory of evolution or intelligent design?

After all, if we believe God created everything, what difference does it make if He created the world instantly or used the process of evolution to develop everything gradually over long ages?

The late Dr. D. James Kennedy reminds us that we cannot have it both ways.

"A naturalistic, evolutionary world view is more than a question of whether evolution has ever taken place on any scale, at any time, in any place. Rather it represents the complete antithesis of the Christian biblical world view. Either the Bible is true when it says a personal God created the heavens and earth and all it contains, or evolution is true. No one can have it both ways."

Chapter 1

Two Competing Philosophies

Ron Carlson said: "One of the most important questions that anyone can ask today is regarding the question of origins.

Today there are two competing philosophies on this issue. One is the theory of evolution, which says that men and women are merely an accident, evolved from slimy algae. The other view is Genesis 1:1, which says, "In the beginning God created.

How you answer the question of origins, whether you are an accident or a unique creation of God, will ultimately determine everything else in your life.

It will determine your value for human individuals, your basis of morality, your meaning and purpose in life, and your ultimate destiny.

It is one of the most fundamental questions you can ask!" [1]

---

[1] Carlson, Ron (2003-07-01). Fast Facts® on False Teachings (Kindle Locations 857-865). Harvest House Publishers. Kindle Edition.

We believe both **true Science** and the Bible support **energy intelligently directed.** (Emphasis ours).

We reject the notion that nothing x time x chance = everything (Emphasis ours).

We reject the "molecules to man" theory of origins.

This book was written to encourage our readers to look at the scientific evidence that points to the fact that you are not an accident, but a unique creation of an almighty loving Creator. We report. You decide.

Chapter 2

What is Evolution?

"The general theory of organic evolution is the theory that all living things have arisen by a materialistic, naturalistic, evolutionary process from a single source which itself arose by a similar process from a dead, inanimate world. This theory may also be called the molecule-to-man theory of evolution." (Duane T. Gish, Evolution: The Fossils Say No! San Diego: Institute for Creation Research: 1972, page 1)

In the opinion of this present writer, there is no conflict between true science and the Bible. There is, however, a great conflict between "true science" and the *theory of evolution*.

Micro-evolution

Micro-evolution is a term, first used in 1911 to describe comparatively *minor evolutionary change* involving the accumulation of variations in populations usually below the species level. For example: There are many different breeds of dogs and many different sizes in finch beaks.

Macro-evolution

Macro-evolution is a term first used in 1939 to describe relatively *large and complex changes*, as in species formation. Evolutionists use this to explain one species evolving into a completely new species (kind or family).

Species / Kind / Family

A species is a group of animals or plants that are similar and can produce young animals or plants. Examples: Dogs, cats, horses, and humans.

We have no problem with **micro-evolution because there** are variations within each species.

We do have a problem this *macro-**evolution*** because it is not scientific. Examples: Cats becoming dogs and horses becoming humans. There is absolutely no scientific evidence for transmutation of species. This is wishful thinking, junk science, has never been seen and is totally refuted by the fossil record.

Charles Darwin

You may or may not be aware of the fact, that Charles Darwin *was not a scientist* (emphasis ours). The only academic degree he held was in theology.

Charles Darwin's father was a medical doctor and his desire was for his son, Charles, to follow in his steps. For some reason, Charles failed to complete his medical training and chose, as an alternate, to prepare for the ministry.

Chapter 3

Evolution is a Sacred Cow

The theory of evolution is so widely accepted today because it is the only theory which can be legally taught in our public-school system.

It is protected by law and the American Civil Liberties Union and other watchdog groups which are ready to burden local school boards with excessive lawsuits, if anything but evolution is taught in the classroom.

There is also much pressure on public schools to stop even referring to evolution as a "theory" and requiring the schools to call evolution a "fact."

Why Do They Teach Evolution?

Dr. Phillip E. Johnson, Professor of Law at the University of California at Berkeley, has written a book exposing the falsehood of evolution entitled "Darwin on Trial."

"Some people have asked us, if there is no evidence for evolution, why do teachers continue to propagate it in our universities and schools?"

Dr. Johnson was speaking at a conference when he was asked this question. His reply was very interesting coming from someone within the academic community:

"Most professors continue to teach evolution in the universities out of fear. This fear is that of not being tenured, of not getting research grants, of not being published, and of not being accepted by their peers.

So, to be accepted, to be published, to be granted research money, and to be tenured by their university, they must follow the party line, which is evolution. This is how the academic game is played." [2]

The True Role of Public-School Teachers

Very *few parents have a clue* as to true objective of todays' public (government) school system.

Perhaps these comments taken from an address at Harvard in 1973 will shed some light on the subject. These remarks were taken from the Humanist Magazine.

---

[2] Carlson, Ron (2003-07-01). Fast Facts® on False Teachings (Kindle Locations 1031-1038). Harvest House Publishers. Kindle Edition.

"The battle for mankind's future must be waged and won in the public- school classroom by teachers who correctly perceive their role as the proselytizers of a new Faith. The classroom must and will become the arena of conflict between the old and the new, The rotting corpse of Christianity and the new faith of humanism."

Do you understand what he said? Let' read the last sentence again: "The classroom must and will become the arena of conflict between the old and the new . . . **The rotting corpse of Christianity and the new faith of humanism."**

Christian Students Targeted

You should understand by now that humanism is the official philosophy/religion of every government school in America, and the theory of evolution is one of the five major doctrines upon which humanism is built.

Students in our public (government) schools who fail to go along with the theory of evolution face unbelievable discrimination from teachers.

In Iowa, one professor stated that the believed he should be allowed to fail any student he discovers believing in creation. The professor desires to fail the student regardless of his/her grade point average.

Another professor suggested that they be allowed to take back student's grades if the professor learns that the student embraces creation even after graduation.

Here are samples of what our students are being taught by the evolutionists. The following quotes are from the late D. Kennedy's book, *Today's Conflict, Tomorrow's Crisis*.

Schoenberg: "Man is merely a "hairless ape."

DuMaurier: "We are a fungus on the surface of one of our minor planets."

F. P. Church: "The most noble individual is a mere insect, an ant."

H. L. Mencken: "The cosmos is a gigantic fly-wheel making 10,000 revolutions a minute. Man is a sick fly taking a dizzy ride on it. Religion is the theory that the wheel was designed and set spinning to give him a ride."

W.B. Thornburry: "Human existence was a jest, a dream, a show, bubble, air."

True Science Versus Evolution

In the opinion of this present writer, there is no conflict between *true science* and the Bible.

There is, however, a great conflict between "true science" and the *theory of evolution*. In the opinion of this writer, evolution is not based on true science.

There is *true science* and there is false science, and there are *scientific facts* that are censored from all public-school textbooks because evolution is a sacred cow in all government schools.

Consider the exhortation from Paul to Timothy in I Timothy 6:20: "O Timothy, keep that which is committed to thy trust, avoiding **profane and vain babblings, and oppositions of science falsely so called**."

The evolutionist tries to slip their theory in under the radar disguised as science. The purpose of this book is to strip away their disguise and expose evolution for what it really is. It is a very weak theory that is not supported by modern science.

Chapter 4

When and where did Life Begin?

This is the picture which appears in Volume 311, Number 3, pages 40-41 of the September 2014 edition of the **Scientific American.** If you want the *latest information* on the theory of evolution, order this copy.

There are basically four theories put forth by the evolutionists to explain how life began on earth. None of these are based on science because nobody was there when it happened. All of them are man-made theories.

They should begin with *"Once upon a Time."* True Science is based on observation and experiment. Evolutionists believe that n*othing x time x chance = everything.*

1. Spontaneous Generation

According to this theory, all life on the earth sprang from dead matter. True scientists understand that Louis Pasteur settled this argument scientifically years ago in his experiments with dead meat, blow flies, and maggots. Pasteur demonstrated that when something is sterile, it is sterile. Life has never evolved from dead matter. This is true science.

2. Life Began Accidentally

According to this theory, life began accidentally, by chance, perhaps when lightning passed through vapors that surrounded the earth and a spark of life fell into the ocean. From that tiny speck evolved all the life forms we see on earth today.

Again, this theory, like the first theory dodges the real question. Tell us, Mr. Evolutionist, where did the lightning came from?

Tell us where the gasses came from.

Tell us where the oceans came from.

We have no objection to theories, but we do have a problem when make believe theories like these are passed off as scientific fact.

Remember! To qualify as science the idea must be verified by observation and experiment.

3. Directed Transpermia

According to the evolutionists, the first tiny germ of life came to the earth on a meteor from outer space.

Sir Francis Crick, co-discoverer of DNA, believes there are life sperms everywhere in space and some of these have been directed by advanced galactic civilizations to the earth.

Both Crick and Leslie Orgel of the University of California in San Diego have made strong statements of the impossibility of life originating on earth by chance. Crick says:

"If a particular amino acid sequence was selected by chance, how rare would that event be?

Suppose the chain is about 200 amino acids long; that is, if anything rather less than the average length of proteins of all types. Since we have just 20 possibilities at each place, the number of possibilities is 20 multiplied by itself 200 times.

That is approximately equal to 1 in 10 followed by 260 zeros.

The great majority of sequences can never have been synthesized at all, at any time." (Sir Francis Crick, Life Itself (New York, NT: Simon and Shuster, 1981,) p. 51. This is also quoted by Henry Morris in The Modern Creation Trilogy (Green Forest, AR, 1996,) Master Books, Inc.) Pages 151-152)

This "germ from space" theory dodges the real questions, which are:

Where did the other planets come from?

Where did the meteor come from?

Where did the germ of life come from?

How did the germ survive the terrific heat when the meteor entered the earth's atmosphere? This is a theory, not science!

4. The Big Bang Theory

Another very popular theory today is known as the big bang theory. When discussing this theory with an evolutionist, we recommend you quote evolutionists, not preachers. Use the evolutionists own words in your discussions with them.

Chet Raymo: Astronomer

Author Phillip Yancy did this in a debate with some noted professors. Yancy quoted Chet Raymo, an astronomer and science writer, who calculated the odds of our universe resulting, as he believes it did, from sheer chance. This famous astronomer said:

"If, one second after the Big Bang, the ratio of the density of the universe to its expansion rate had differed from its assumed value by only one part in 10 15th (10 followed by 15 zeros), the universe would have either quickly collapsed upon itself or ballooned so rapidly that stars and galaxies could not have been condensed from the primal mater."(Phillip Yancy, Soul Survivor. (New York: Doubleday) 2001, page 3

The evolutionists would have us believe that nothing x time x chance = everything. Is this scientific?

The fact is, there is not one scientifically observed and recorded example of spontaneous generation in history. Life produces life. Life never came from dead matter.

Sir Fred Hoyle

What are the odds for just one amino acid coming into existence by chance? Amino acids are the molecular units that make up proteins. All proteins are various compositions of *twenty specific* naturally occurring amino acids.

Sir Fred Hoyle, former professor of Cambridge and discoverer of steady state cosmogony decided to find out the probability of a simple cell coming into existence anywhere in the universe.

Hoyle calculated that the odds of even one amino acid coming into existence by chance were 1 in 10 to the 40th power. **That's one in ten with 40 zeros behind it.**

We must remind the Evolutionists from time to time that for something to qualify as "science" it must be based on observation and experiment. Evolution doesn't pass this test.

Chapter 5

Why Darwin Rejected the Fossil Record

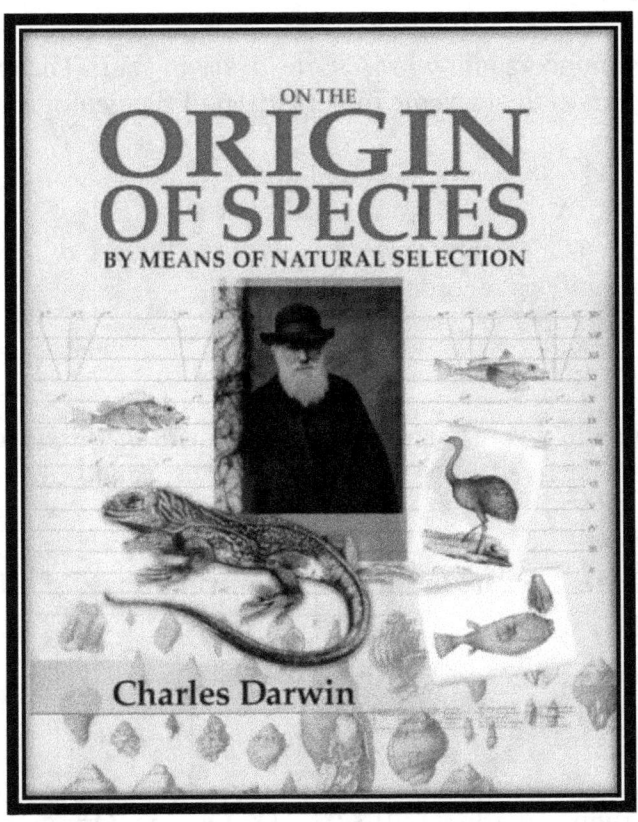

When confronted by the fact that there is no evidence of evolution in the fossil record, Charles Darwin gave the following feeble explanation:

"Why then is not every geological formation and every stratum full of such intermediate links? Geology assuredly does not reveal any such finely graduated organic chain; and this, perhaps, is the most obvious and serious objection which can be urged against the theory. The explanation lies, as I believe, in the <u>extreme imperfection of the geological</u> record." [3]

Did you understand what Charles Darwin said? The explanation lies, however, in the <u>extreme imperfection</u> of the geologic record."

Geology

Geology is a science that *deals with the history of the earth* and its life especially as recorded in the rocks.

Paleontology

Paleontology is the science *that deals with the fossils* of animals and plants that lived very long time ago, especially in the time of dinosaurs.

There is No Life in the Pre-Cambrian Layer. Period!

———————————

[3]   Darwin, Charles (2013-03-21). On the Origin of Species (Kindle Locations 5235-5238). Arcturus Publishing. Kindle Edition.

Cambrian Layer

When we come to the Cambrian layer, not only do we find life, we observe that *life is already diversified and complex.* There are no transitional forms. There are no "missing links." In fact, the entire "chain" is missing.

Ants Trapped in Amber

Another fatal blow to the evolutionist, is the fact that there are ants trapped in amber, that evolutionist believe to be millions of years old. To the evolutionist dismay, the ants trapped in amber are just like their cousins that are found in this writer's sugar bowl in Newberry, Florida in 2014

Variations

There are all sorts of variations within a specie (kind or family), but there is no transmutation of species.

There are all sorts of dogs, all sorts of cats, and all sorts of horses. They come in all sizes and colors, but they are still "locked in" their "kind."

Dr. Duane Gish

"As Dr. Duane Gish, who received his Ph.D. in biochemistry from the University of California at Berkeley, pointed out in Acts and Facts (published by the Institute of Creation Research, El Cajon, CA):

"The fossil record shows the sudden appearance, fully formed, of all the complex invertebrates (snails, clams, jellyfish, sponges, worms, sea urchins, brachiopods, and trilobites) without a trace of ancestors. He goes on to add:

The fossil record also shows the sudden appearance, fully formed, of every major kind of fish (supposedly the first vertebrates) without a trace of ancestors.

This proves beyond a reasonable doubt that evolution has not occurred. If evolution has occurred, our museums should contain thousands of fossils of intermediate forms. However, not a trace of an ancestor or transitional form has ever been found for any of these creatures!

To sum it up: We have never observed evolution in the fossil record, and we have never observed evolution in the natural world. Evolution is a theory that exists only in the imaginations of evolutionists." (Carlson, Ron (2003-07-01). Fast Facts® on False Teachings (Kindle Locations 1002-1013). Harvest House Publishers. Kindle Edition.)

Dr. Henry Morris

Dr. Henry Morris reminds us: "Actually, there is no evidence at all that evolution ever took place in the past either. In all recorded history, extending back nearly five thousand years, no one has ever recorded the natural evolution of any kind of creature (living or non-living) into a more complex kind. Furthermore, all known vertical changes seem to go in the wrong direction.

An average of at least one species has become extinct every day since records have been kept, **but no new species have evolved during that time** (Emphasis ours).

Again, however, the story is one of extinction, not evolution.

Numerous kinds of extinct animals are found (e.g., dinosaurs), but never, in all these billions of fossils, is a truly incipient or transitional form found. No fossil has ever been found with half scales/half feathers, half legs/half wings, half-developed heart, half-developed eye, or any other such thing.

If evolution were true, there should be millions of transitional types among these multiplied billions of fossils. In fact, everything should show transitional features. But they do not! If one were to rely strictly on the observed evidence, he would have to agree that past evolution has also been falsified."

Summary

There are no transitional forms in the fossil record.

What do we find in the fossil record? We discover that life appeared suddenly and when it appeared, it was already diversified and complex.

The theory of evolution is based on "historical" science, not "true" science.

And yet, despite all of this, evolution is the only view that can legally be taught in our public (government) schools.

Chapter 6

True Science versus Historical Science

There is not one *scientifically observed and* recorded case of transmutation of species (kinds/families) in the history of science.

Evolution is not scientific. It is a weak theory accepted by blind faith. The formula: n*othing x time x chance = everything,* is not scientific. It's just wishful thinking.

There's true Science and there is "junk" science. True Science has always pointed to an intelligent Creator. Wherever one looks in this universe we can see intelligent design.

There are hundreds of reputable scientists who do not believe in evolution.

Whether we look through a telescope or a microscope we can see intelligent design. There are no outlaw spaces anywhere in the universe.

The only people who don't see this are people who refuse to see this because they have closed their minds.

Evolution is not happening today. It was never observed by anyone, at any time in the recorded history of man. It is a theory based on blind faith. It is widely accepted today because it is the only explanation students are allowed by law to hear.

True science is based observation and experiment.

Evolutionist can't tell us when, where, or how life began. Their theories require too much faith for this present writer to accept.

Spontaneous generation is not scientific. It's accepted by blind faith.

The fossil record reveals that life appeared suddenly and when it appeared it was already diversified and complex. There are no transitional forms in the fossil record.

Intelligent design makes sense. Spontaneous generation does not make sense.

It is not intelligent to believe that life came into existence out of nothing or by blind chance.

Where in the history of the world has anything come into existence without energy intelligently directed?

Have you ever seen a master painting without a painter?

Have you seen a large building or a bridge without an engineer?

Look at the human eye. Look at the hummingbird. Look at DNA.

What marvelous, complex examples of intelligent design.

Louis Pasteur

Louis Pasteur proved scientifically that life did not come from dead matter. This is a fact of true science. Junk science tells us that nothing x time x chance = everything.

No reputable scientist can point to even one scientifically observed and recorded instance of anything coming into existence out of a vacuum or coming into existence without energy intelligently directed.

Radiation and Drosophila

Changes must take place genetically. A prime example of this is drosophila (Fruit Fly). "The fruit fly has long been a favorite object of mutation experiments because of its fast gestation period (12 days). X- Rays have been used to increase the mutation rate in the fruit fly by 15,000 percent.

All in all, scientists have been able to catalyze the fruit fly evolutionary process such that what has seen to occur in drosophila is the equivalent of many millions of years of normal mutations and evolution. Even with this tremendous speedup of mutations, scientist have never been able to come up with anything other than another Fruit Fly. "

Thermodynamics

The Second Law of Thermodynamics (entropy) deals with the conservation of energy. This is a law of science. The second law wipes out the theory of evolution.

Evolution is change outward and upward. Entropy is change inward and downward. Everything in the universe loses energy. Everything runs down. Evolution teaches the opposite.

The law of increasing entropy is an impenetrable barrier which no evolutionary mechanism has ever been able to overcome.

Henry Morris reminds us: "Stars explode, comets and meteorites disintegrate, the biosphere deteriorates, and everything eventually dies, so far as all historical observations go, but nothing has ever evolved into higher complexity."

Mathematical Laws of Probability

It takes way too much blind faith to become an atheist and to believe that nothing x time x chance = everything.

Creation requires a Creator. It is not mathematically probable that life began by chance. Every honest scientist knows this.

No matter where you look in this universe, you see evidence that there is a Creator.

Creation speaks of a Creator

Everywhere you look, you see order and design.

Whether we look through a microscope or through a telescope, we all see the same thing. We see order and design.

Someone said that one of the greatest scientific discoveries in all time was that there were no outlaw spaces anywhere in the universe.

Recently someone sent me an interesting Email which illustrates this point. If I could remember who sent it, I would give them credit.

Hatching Eggs

God's order may be observed in the hatching of eggs. For example:

Eggs of the potato bug hatch in 7 days

Eggs of the canary in 14 days

Eggs of the barnyard hen hatch in 21 days

Eggs of ducks and geese hatch in 28 days

Eggs of the mallard hatch in 35 days

The eggs of the parrot and the ostrich hatch in 42 days.

Did you notice that all divisible by seven, the number of days in a week?

Fruits, Grains, & Vegetables

The Email containing this information continued: "God's wisdom is revealed in His arrangement of fruit, vegetable, and grain seed sections and segments, as well as in the number of seeds/grains.

Each watermelon has an even number of stripes on the rind.

Each orange has an even number of segments.

Each ear of corn has an even number of rows.

Each stalk of wheat has an even number of grains.

Every bunch of bananas has on its lowest row an even number of bananas, and each row decreases by one, so that one row has an even number and the next row an odd number.

The waves of the sea roll in on shore twenty-six to the minute in all kinds of weather.

All grains are found in even numbers on the stalks, and the Lord specified thirty-fold, sixty-fold, and a hundredfold - all even numbers.

The Email reminded us: "God has caused the flowers to blossom at certain specified times during the day, so that Linnaeus, the great botanist, once said that if he had a conservatory containing the right kind of soil, moisture and temperature, he could tell the time of day or night by the flowers that were open and those that were closed!"

Acquired Characteristics are Not Inherited

Evolutionists have long taught that giraffe's have long necks because their ancestors stretched their necks to reach the leaves on the trees and this characteristic was passed down to their offspring. No real scientist believes that today.

Gregor Mendel established years ago by his study of genetics and the laws of heredity, that changes that are inherited must take place in the genes. What happens if we cut off the tail of a mother dog and she has puppies? Will her puppies have bobbed tails? Acquired characteristics cannot be inherited. This is a fact of science

If acquired characteristics could be inherited, the children of mailmen would have huge legs. The children of college professors would have eyes as large as fishbowls, resulting from spending long hours reading.

Chapter 7

Cave Men and Dinosaurs in the Bible

This picture appeared in the September 2014 issue of the *Scientific American*. According to evolutionist, this man ate grass. The fact is, there were cave men in Jobs' day who ate grass. See Job 30:4.

Cave Men (Job 30:1-8)

1 But now they that are younger than I have me in derision, whose fathers I would have disdained to have set with the dogs of my flock.

2 Yea, whereto might the strength of their hands profit me, in whom old age was perished?

3 For want and famine they were solitary; fleeing into the wilderness in former time desolate and waste.

4 Who cut up mallows by the <u>bushes, and juniper roots</u> for their meat (emphasis ours).

5 They were driven forth from among men, (they cried after them as after a thief;)

6 To dwell in the <u>cliffs of the valleys, in caves</u> of the earth, and in the rocks.

7 Among the bushes they brayed; under the nettles they were gathered together.

8 They were children of fools, yea, children of base men: they were viler than the earth.

There is no question of "cave men." The fact is, they lived at the same time other normal people lived.

Dinosaurs (Job 40:15-24)

15 Behold now **behemoth**, which I made with thee; he eateth grass as an ox.

16 Lo now, his strength is in his loins, and his force is in the navel of his belly.

17 He moveth his tail like a cedar: the sinews of his stones are wrapped together.

18 His bones are as strong pieces of brass; his bones are like bars of iron.

Could Job be Referring to Apatosaurus?

19 He is the chief of the ways of God: he that made him can make his sword to approach unto him.

20 Surely the mountains bring him forth food, where all the beasts of the field play.

21 He lieth under the shady trees, in the covert of the reed, and fens.22 The shady trees cover him with their shadow; the willows of the brook compass him about.

23 Behold, he drinketh up a river, and hasteth not: he trusteth that he can draw up Jordan into his mouth.

24 He taketh it with his eyes: his nose pierceth through snares.

COMPARE THIS TO THE EVOLUTIONIST THEORIES OF CAVE MEN

Pithecanthropus Erectus (Java Man)

The following material is quoted from the book "Did Man Just Happen," by the late Dr. W.A. Criswell. PDF copies of these messages are available @ **www.Criswell.com**

"In 1891, Dr. Eugene Dubois, a medical doctor in the Dutch army, and an ardent evolutionist, went to Java to find the "missing link." The good doctor found a piece of a skull cap and a piece of a left thigh bone.

The fragments were not found in the same place or the same time. Neither were they found in stone (like fossils). They were found approximately 50-75 feet apart, in the sands of a riverbed, over a period of one year.

Notice what the "scientists" said about the discovery of Pithecanthropus Erectus:

'It is fortunate that the most distinctive portions of the human frame have been preserved, because from these specimens, we were able to reconstruct the entire being. This man stood half-way between the anthropoid and the existing man.'

What did the good doctor find that qualified them as the most distinctive parts? He found a piece of a skull cap and a fragment of the left knee bone. And from these fragments, the evolutionists constructed the entire man.

Hesperopithecus Haroldcookii (The Nebraska man)

According to Dr. Criswell, Hesperos means evening or Western.

Pithecus is the Greek word for ape.

Haroldcookii is a scientific sounding way to say, "Harold Cook," the name of the man who discovered the remains of the Nebraska Man.

The London Illustrated Times featured a picture of the Nebraska man and his wife.

http://bevets.com/nebraska.htm

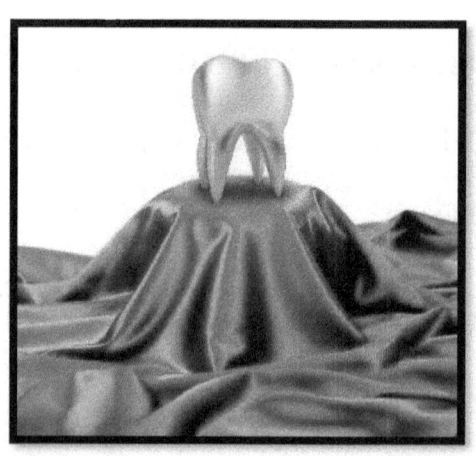

We might add that the Nebraska Man was the great scientific evidence used by evolutionist in the famous Scopes trial in Dayton, Tennessee, in 1925. What modern textbooks don't tell students today is what Harold Cook really found. What did he find? **Harold Cook found one tooth! Not teeth! He found one tooth**, and that tooth was later found to be that of an extinct species of pigs. From that one tooth, Evolutionists took plaster of Paris and fashioned the man and his wife. So much for the "scientific evidence" for evolution.

Eanthropus Dasani (The Piltdown Man)

The Piltdown man skull of 1912 was a fraud. This skull was a replica of Dawson's so-called "Dawn Man," which had been unearthed in a gravel pit at Piltdown near Sussex, England, by Charles Dawson in 1912.

The find consisted of a modern appearing cranium, with a modern sized brain, combined with a primitive apelike jaw, and was found near the teeth of extinct animals dated at five million years old.

For 40 years the skull bones of "Dawn Man" were considered genuine and hindered understanding of human evolution by supporting the biased view that a large brain led the evolutionary way toward modern humans.

By 1953, the application of fluorine analysis and the work of two anatomists and an archeologist exposed *Piltdown Man* as a hoax.

The "find" turned out to be a modern appearing human cranium and the mandible of a modern orangutan, buried along with the bones of the extinct animals.

The point is, both the Bible and true science speak of cave men. Job spoke of these cave men eating grass. This is true Science.

Historical science (once upon a time science) teaches that the cave men who ate meat evolved millions of years later than the cave men who age grass. This is an outright fabrication.

There were cave men and dinosaurs living on earth in the days of Job. Job had seen them and wrote about them.

Summary

All the Evolutionist needs is blind faith and plaster of Paris. They can take a pigs' tooth and create a man and his wife and place them in a museum. Nobody questions this because evolution is a sacred cow.

The problem is, the theory of evolution is the only concept students in public schools are allowed to hear. It is a closed system and neither students nor their teachers have been allowed to even be exposed to anything but that. That's why there's so much confusion on this subject.

Chapter 8

The Sad Legacy of Charles Darwin

Evolution Destroys Self Esteem

For an example of the effect of the theory of evolution on a persons' self-esteem, consider the following quotes from the book "Fast Facts on False Teaching" by Ron Carlson.

According to the evolutionists, you are the descendant of a tiny cell of primordial protoplasm that washed up on an ocean beach 3½ billion years ago.

"You are the blind and arbitrary product of time, chance, and natural forces. Your closest living relatives swing from trees and eat crackers at the zoo. You are a mere grab bag of atomic particles, a conglomeration of genetic substance. You exist on a tiny planet in a minute solar system in an obscure galaxy in a remote and empty corner of a vast, cold, and meaningless universe. You are flying through lifeless space with no purpose, no direction, no control, and no destiny but final destruction. You are a purely biological entity, different only in degree but not in kind from a microbe, virus, or amoeba. You have no essence beyond your body, and at death you will cease to exist entirely. In short, you came from nothing, you are going nowhere, and you will end your brief cosmic journey beneath six feet of dirt, where all that is you will become food for bacteria and rot with worms." [4]

---

[4] Carlson, Ron (2003-07-01). Fast Facts® on False Teachings (Kindle Locations 1077-1079). Harvest House Publishers. Kindle Edition.

Evolution Motivated Adolph Hitler

The ideas of Darwin even paved the way for Hitler, who used them to justify the extermination of those he considered less than ideal resulting in the mass murder of millions of Jews, gypsies, and others. His ideas have contributed to the erosion of the family, educational institutions, the decay of the legal system, and have led to great compromise in the Church."

(Ham, Steve; Ken Ham (2008-08-31). Raising Godly Children in an Ungodly World (p. 28). Master Books. Kindle Edition.)

The Church Sang Louder

The church did nothing. Hitler told the religious leaders to take care of religion and he would take care of politics. They did and he did.

When the trains filled with Jews on their way to the gas chambers would pass by the churches, the people in the churches would sing louder until the train passed so they wouldn't hear the screams of those on their way to the death chambers.

Evolution and Planned Parenthood

For some interesting information on the impact of the theory of evolution on Planned Parenthood and abortion, we recommend you go on Google and type in *"10 Eye-opening Quoted from Planned Parenthood Founder."*

Margaret Sanger, "Birth Control and **Racial** Betterment - The Public ...
www.nyu.edu/projects/**sanger**/webedition/app/.../show.php?**sanger**Doc...
Margaret Sanger, "Birth Control and **Racial** Betterment," Feb 1919. Published article.
Source: Birth Control Review, Feb. 1919. , Library of Congress Microfilm ...
You visited this page on 10/11/14.

10-Eye-Opening Quotes From Planned Parenthood Founder ...
www.lifenews.com/.../10-eye-opening-quotes-from-planned-parenthood-fo...
Mar 11, 2013 - **Margaret Sanger** has been lauded by some as a woman of valor, but a ... had
some unsavory things to say about matters of **race**, birth control, ...
You've visited this page 2 times. Last visit: 10/11/14

Margaret Sanger, Founder of Planned Parenthood, In Her Own Words
www.dianedew.com/**sanger**.htm
Margaret Sanger, Pivot of Civilization, referring to immigrants and poor people. On
sterilization & **racial** purification: Sanger believed that, for the purpose of ...
You visited this page on 10/11/14.

BlackGenocide.org | The Negro Project
www.blackgenocide.org/negro.html
The NEGRO PROJECT: **Margaret Sanger's** EUGENIC Plan for Black America ... economic
empowerment, elevate the **race** and garner the respect of whites.
You visited this page on 10/11/14.

Evolution Promotes Racism

This Picture Offends Me

This picture is offensive to me. It shows two men. One is a black man and the other is a white man.

The reason this picture is offensive to me as a Christian is because the theory of evolution clearly teaches that black people are lower in the Evolutionary scale than white people.

"The implications of Darwin's legacy are far-reaching. He paved the way for moral relativism, and fueled racism claiming that blacks, aborigines, and others are inferior, less-evolved races."

The Impact of Evolution on our Churches

Pastors and church workers need to read Ken Ham's book, *Already Gone*. Those who understand what's really going on already know that from 60-80% of the young people in our churches will leave church after graduation and never return. Statistics reveal that their bodies are still in church until graduation, but their minds are already gone.

60 % are Already Gone

"Based on interviews with 22,000 adults and over 2,000 teenagers in 25 separate surveys, George Barna unquestionably quantified the seriousness of the situation: six out of ten 20-somethings who were involved in a church during their teen years are already gone."[5]

Baptists lose 70-88%

According to a survey by the Southern Baptist Convention, Baptist churches lose more than other denominations. According to the survey, Baptists lose from 70-88% of their young people. They will leave church after graduation and never come back.

Sunday School has also Failed

"This was our most stunning and disconcerting result of the entire survey.

---

[5] Hillard, Todd; Britt Beemer; Ken Ham (2009-05-01). Already Gone (Kindle Locations 210-212). Master Books. Kindle Edition.

First, we found out that we were losing our kids in elementary school, middle school, and high school rather than in college.

Then we found out that Sunday school is one of the reasons why. The 'Sunday school syndrome' is contributing to the epidemic, rather than helping alleviate it.' "

These numbers are statistically significant and contrary to what we would expect. This is a brutal wake-up call for the Church, showing how our programs and our approaches to Christian education are failing dismally. [6]

The obvious conclusion is that Sunday school really had no impact on what children believed in these critical areas.

"For example, when asked if they believed in the **creation of Adam in the Garden of Eden**, Sunday school had no significant effect on the answers.

The same can be said for the story of **Sodom and Gomorrah** and Lot's wife.

The same can be said of Noah's ark and the global Flood.

Belief in the **Tower of Babel** was nearly identical.

---

[6] Hillard, Todd; Britt Beemer; Ken Ham (2009-05-01). Already Gone (Kindle Locations 410-415). Master Books. Kindle Edition).

In these areas Sunday school did nothing – it wasn't a help or a detriment.

The numbers indicate that Sunday school didn't do anything to help them develop a Christian worldview.

In several other areas, as shocking as this sounds, the reality we have to face is that Sunday school clearly harmed the spiritual growth of the kids." [7]

The Church Has Lost this War

Pastors and church members who are still dreaming of salvaging our public-school system are like those pitiful Japanese soldiers found on isolated islands after World War II.

They didn't even know the war was over. Japan lost. America won.

When it comes to the battle for the minds of our children, the war is over. The church lost.

---

[7] (Hillard, Todd; Britt Beemer; Ken Ham (2009-05-01). Already Gone (Kindle Locations 431-437). Master Books. Kindle Edition.

Chapter 9

The Legacy of Bible Believing Christians

Are you aware that Bible-believing Christians built the first schools America? Christians built nearly all the colleges and universities in America. Christians built Harvard in 1638, Yale in 1701, Princeton in 1746, and Dartmouth in 1754.

The original mission of the first American colleges was to train men for the ministry. In the 17th century, 52% of the men who graduated from Harvard entered the ministry.

In 1646, **Harvard adopted Rules & Precepts**. Read this carefully and compare this to what is being taught at Harvard today.

"Everyone shall consider that the main end of his life and studies is to know God and Jesus Christ which is eternal life." [8]

---

[8] (The Rebirth of America, Philadelphia: Arthur DeMoss Foundation, 1986, page 41)

Dr. Henry Morris

Consider the following list of born-again, Bible-believing scientists and the fields of study for which they are famous.

This impressive list is found on pages 463-464 in Henry Morris' book titled, "The Biblical Basis for Modern Science."

"Antiseptic Surgery - Joseph Lister

Bacteriology - -Louis Pasteur

Calculus - Isaac Newton

Celestial Mathematics - Johann Kepler

Chemistry - Robert Boyle

Comparative Anatomy - Georges Cuvier

Computer Science - Charles Babbage

Dimensional Analysis -Lord Rayleigh

Dynamics - Isaac Newton

Electrodynamics - James Clark Maxwell

Electromagnetics - Michael Faraday

Electronics - Ambrose Fleming

Energetics - Lord Kelvin

Entomology of living insects - Henri Fabre

Field Theory - Michael Faraday

Fluid Mechanics - George Stokes

Gas dynamics - Robert Boyle

Genetics - Gregor Mendel

Glacial Geology - Louis Agassiz

Gynecology - James Simpson

Hydraulics - Leonardo da Vinci

Hydrography - Matthew Maury

Hydrostatics - Blaise Pascal

Ichthyology - Louis Agassiz

Isotopic Chemistry - William Ramsey

Model Analysis - Lord Rayleigh

Natural History - John Ray

Non-Euclidean Geometry - Bernhard Riemann

Oceanography - Matthew Maury

Optical Mineralogy - David Brewster

Paleontology - John Woodard

Pathology - Rudolph Virchow

Physical Astronomy - Johann Kepler

Reversible Thermodynamics - James Joule

Statistical Thermodynamics - James Clark Maxwell

Stratigraphy - Nicholas Steno-

Systematic Biology – Carolus Lennaeus

Thermodynamics - Lord Kelvin

Thermokinetics - Humphry Davy

Vertebrate Paleontology - Georges Cuvier

Absolute Temperature scale - Lord Kelvin

Actuarial Table - Charles Babbage

Barometer - Blaise Pascal

Biogenesis Law - Louis Pasteur

Calculating Machine - Charles Babbage

Chloroform - James Simpson

Classification System - Carolus Linnaeus

Double Stars - William Herschel

Electric Generator - Michael Faraday

Electric Motor - Joseph Henry

Ephemeris Table - Johann Kepler

Fermentation control - Louis Pasteur

Galvanometer - Joseph Henry

Global Star catalogue - John Herschel

Inert Gases - William Ramsey

Kaleidoscope - David Brewster

Law of Gravity - Isaac Newton

Mine Safety Lamp - Humphry Davy

Pasteurization - Louis Pasteur

Reflecting telescope - Isaac Newton

Scientific Method - Francis Bacon

Self-induction - Joseph Henry

Telegraph - Samuel B. Morse

5 Thermionic Valve - Ambrose Fleming

Trans-Atlantic cable - Lord Kelvin

Vaccination and Immunization - Louis Pasteur

Electrodynamics - James Clerk Maxwell

Galactic Astronomy - William Herschel" [9]

In addition to all of all of these, Bible-believing Christians helped abolish slavery, abolished child labor, established hospitals, and hundreds of humanitarian projects.

---

[9] (Henry Morris, "The Biblical Basis for Modern Science." Pages 463-464.)

Chapter 10

Reason with your mind - not your Heart

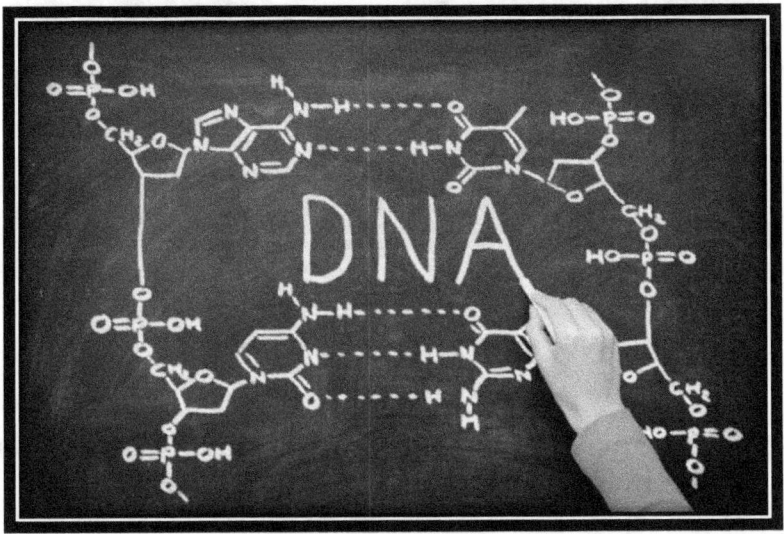

A few years ago, this writer came across an article in the Reader's Digest titled, *Ten Reasons Why Scientists Believe in God.* This timely article approached the existence of God from the position that there were so many exacting conditions necessary for life on earth, that such a system could not have possibly come into existence by chance.

1. Do Satellites and Orbits Point to Blind Chance or Intelligent Design?

We know that if we place a satellite in space, 110 miles from the earth, traveling 17,000 mph it will remain in orbit.

2. Do Days and Nights, based on the Earth's Rotation on its Axis Point to Blind Chance or Intelligent Design?

The earth rotates on its axis at a rate of 1000 mph. If it turned only 100 mph the days and nights would be 10 times as long and life on earth would cease to exist. The days would be so long the sun would burn everything up. What was left would freeze because the nights would also be 10 times

3. Does the Earth's Orbit around the Sun point to Blind Chance or to Intelligent Design?

The earth orbits the sun at a rate of 18.5 miles per second. If the earth slowed down to only 14 miles per second, the earth's orbit would move in so close to the sun that life on earth would be annihilated. If the rate was increased to 26 miles per second, the earth would fly off into outer space and never return.

4. Does the Earth's Tilt at exactly 23 ½ degrees, which causes the Seasons, point to Blind Chance or Intelligent Design?

The earth is tilted at exactly 23.5 degrees. As the earth travels around the sun, this causes the seasons to occur. If it were not tilted this way, the vapors from the ocean would move north and south and pile up on the continents.

5. Does the Earth's Crust suggest Blind Chance or Intelligent Design?

If the crust of the earth was just ten feet thicker, there would be no oxygen and animal life would die out. If the oceans were a few feet higher, carbon dioxide would be absorbed, and plant life could not exist.

6. Does the Distance from the Earth to the Moon suggest Blind Chance or Intelligent Design?

The moon is in an orbit 240,000 miles from the earth. If the moon were just 50,000 miles from the earth, twice each day, ocean tides would rise so high they would cover the continents completely.

7. Do the Spawning Salmon point to Blind Chance or Intelligent Design?

The salmon spend their lives in the ocean. When the time comes to spawn, these fish swim back to the very streams they were hatched in. They will even swim back up the very same side of the river they swam down when they would young.

8. Does the Mystery of the Eels point to Blind Chance or Intelligent Design?

Eels from all over the world swim to the deep ocean off Bermuda to spawn. When those young eels are born, those from Europe return to Europe, and those from America return to America. They never get mixed up or confused. Does this appear to be the result of blind chance or intelligent design?

9. Does Bird Migration point to Blind Chance or Intelligent Design?

The pacific golden plover hatches their young in frozen wastelands of Alaska and the Arctic.

When the young birds are old enough to fly, they fly straight to the islands of Hawaii. They do this without radios, signs, compasses, or a GPS. How do they navigate and find their way? Use your mind! Does this appear to be the result of blind chance or intelligent design?

10. Does the Microscope reveal evidence of Bind Chance or Intelligent Design?

If we look through a microscope, we can see intelligent design.

If we look through a telescope, we can see intelligent design. There are no outlaw spaces anywhere in the universe. The only people who don't see this are people who refuse to see this because they have closed their minds.

The following information and illustrations were taken from a DVD by Louie Giglio. The name of the DVD was "How Great is our God," and it was produced by Passion Conferences.

Have You ever Heard of Laminin?

Laminin is the glue that holds our cells together. Laminin, the tiny protein that holds our cells together is shaped like the cross of Christ. Here's what laminin looks like.

http://www.prayforourleaders.net/index3.html

11. Does the Telescope support Blind Chance or Intelligent Design?

Louie Giglio showed us a picture taken by the Hubble Telescope stationed miles out into space. (You can see this picture on Google). This is a picture of the "Whirlpool Galaxy," located 31 million light years away.

What does that galaxy look like to you? Like the picture of laminin, seen through a microscope, the Whirlpool Galaxy looks like the cross of Christ.

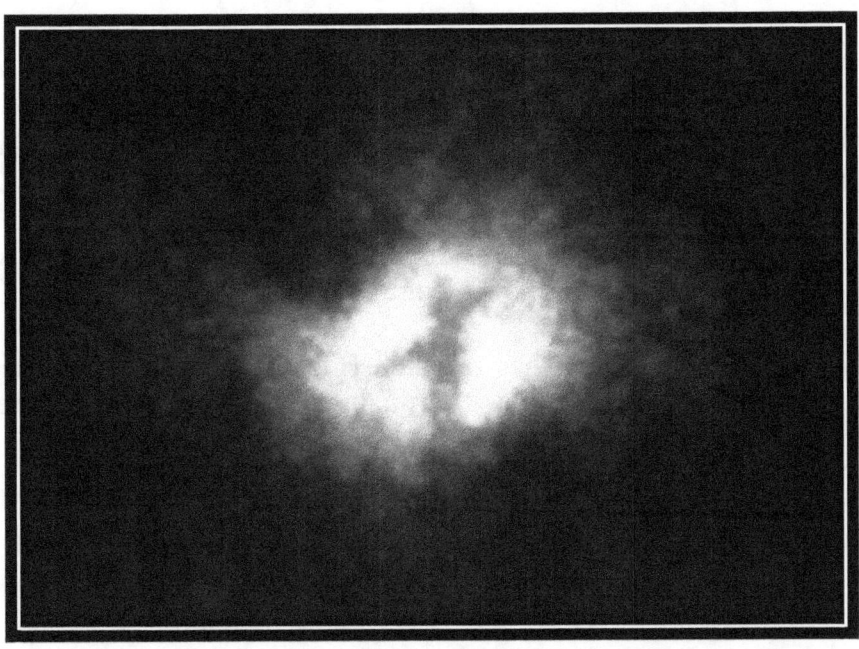

http://9words.ca/2012/04/love-that-shines-through-the-dark/

The God who created laminin and the Whirlpool Galaxy created you. This God not only knows your name, He loves you and has a wonderful plan for your life.

This God also knows that you have sinned, and He came up with a way to forgive you without compromising His Holiness.

If you will turn from your sins, receive Jesus Christ, God's Son, into your life, He will forgive all your sins and adopt you into His own forever family. What about it?

Summary

Use your brain! Do all these things appear to be the result nothing x time x chance = everything? Only a fool would come to that conclusion.

There is not a man of woman on this planet who has an excuse not to believe in God, to acknowledge God and to worship Him.

Why? Because God has revealed Himself through the created universe and has placed a tiny "spiritual receiver" in our consciences so that we know that there is a God.

Chapter 11

A Challenge for Honest Atheists

Atheists Have No Excuse

There is not a person on this earth that has a legitimate excuse for not knowing and worshiping God. No person on earth was ever born an atheist.

There was a definite point in time when every atheist *CHOSEN* to reject the idea of God and become an atheist.

Not only did they choose to become an atheist, they chose to become an atheist for moral reasons, not intellectual reasons. Here's how it works.

God has revealed Himself to every person born on this planet in the history of man. God has given every man and woman a two-fold revelation.

There is the inward revelation and there is the outward revelation. Pascal said: "There is a God-shaped vacuum in every heart." We are all born with an awareness of God.

In addition to that, we can look at the universe around us and we can see evidence of God. We may not know His Name or what He looks like, but we know down in our hearts that there is a God.

Nobody is born an Atheist

The Bible teaches that there is a God and this God has revealed Himself to every person born on this planet. Romans 1 makes it clear that God has given a two-fold revelation to every living person.

Two-fold Revelation

God has revealed Himself through the created, visible universe and He has also revealed Himself in each individual 'heart. We are all born with that awareness.

Romans 1:19-20

"19 For the truth about God is known to them. God has put this knowledge in their hearts.

20 Since earliest times men have seen the earth and sky and all God made and have known of his existence and great eternal power. So, they will have no excuse when they stand before God at Judgment Day."

(Inc. Tyndale House Publishers (2012-05-25). The Living Bible (Kindle Locations 41381-41383). Tyndale House Publishers. Kindle Edition.)

Two Possible Responses

Everyone will respond to this two-fold revelation in one of two ways: They will *accept* the truth, or they will *reject* the truth. If an individual accepts this revelation and responds positively, God will reveal more of Himself to that person and eventually come to know God personally.

If that person responds negatively by rejecting God's revelation, God "turns out their moral and intellectual lights." This condition is called a "reprobate" mind.

If we reject this revelation God will respond by turning out our lights. He allows us to deliberately believe lies. God then washes His hands of us and turns us over to a reprobate (unqualified) mind and this deliberate rejection of Divine revelation opens the door for a chain reaction of all sorts of mental and spiritual degeneration.

Romans 1: 21--28

21 Yes, they knew about him all right, but they wouldn't admit it or worship him or even thank him for all his daily care. And after a while they began to think up silly ideas of what God was like and what he wanted them to do. The result was that their foolish minds became dark and confused.

22 Claiming themselves to be wise without God, they became utter fools instead.

23 And then, instead of worshiping the glorious, ever-living God, they took wood and stone and made idols for themselves, carving them to look like mere birds and animals and snakes and puny* men.

24 So God let them go ahead into every sort of sex sin, and do whatever they wanted to—yes, vile and sinful things with each other's bodies.

25 Instead of believing what they knew was the truth about God, they deliberately chose to believe lies. So, they prayed to the things God made, but wouldn't obey the blessed God who made these things.

26 That is why God let go of them and let them do all these evil things, so that even their women turned against God's natural plan for them and indulged in sex sin with each other.

27 And the men, instead of having normal sex relationships with women, burned with lust for each other, men doing shameful things with other men and, as a result, getting paid within their own souls with the penalty they so richly deserved.

28 So it was that when they gave God up and would not even acknowledge him, God gave them up to doing everything their evil minds could think of."

(Inc. Tyndale House Publishers (2012-05-25). The Living Bible (Kindle Locations 41384-41399). Tyndale House Publishers. Kindle Edition).

Atheists Reason With their Emotions

The Bible says, "The fool has said <u>in his heart</u>, there is no God."

Notice it says that the atheist said in *HIS HEART*, there is no God. What does this mean? That means the Atheists reason with their emotions rather than with their intellects.

Much Do You Know?

If you are an atheist, I would like to ask you a question to illustrate how atheists are basing their position on <u>what they don't know</u>, rather than <u>what they do know</u>.

Try this. Draw a circle on a piece of paper. Let that circle represent all the knowledge and information available in the entire universe.

The question is: What percent of all the knowledge available in the universe to you have?

Do you possess 100% of the knowledge and information available in the entire universe?

Perhaps you have 50%, 25%, or 10% of all available information and knowledge available in the universe?

Let's stretch our imagination and assume you know 25% of everything that be known in this universe. Nobody really knows that much, but let's pretend that you do.

Can you, an Atheist, who knows only 25% of everything that can be known in the universe, be sure God does not exist in the other 75% of the knowledge you do not possess?

Is there one place in the universe where you've not been?

Is there one person you've never met?

Is there anything you do not know?

Is it reasonable to believe that God cannot exist in the 75% of the knowledge in the universe you do not possess? Atheism takes a lot of blind FAITH!

Read Romans 1:18-32 Again

18 For the wrath of God is revealed from heaven against all ungodliness and unrighteousness of men, who hold the truth in unrighteousness;

19 Because that which may be known of God is manifest in them; for God hath shewed it unto them.

20 For the invisible things of him from the creation of the world are clearly seen, being understood by the things that are made, even his eternal power and Godhead; so that they are without excuse:

21 Because that, when they knew God, they glorified him not as God, neither were thankful; but became vain in their imaginations, and their foolish heart was darkened.

22 Professing themselves to be wise, they became fools,

23 And changed the glory of the un-corruptible God into an image made like to corruptible man, and to birds, and four-footed beasts, and creeping things.

24 Wherefore God also gave them up to uncleanness through the lusts of their own hearts, to dishonor their own bodies between themselves:

25 Who changed the truth of God into a lie and worshipped and served the creature more than the Creator, who is blessed forever. Amen.

26 For this cause God gave them up unto vile affections: for even their women did change the natural use into that which is against nature:

27 And likewise also the men, leaving the natural use of the woman, burned in their lust one toward another; men with men working that which is unseemly, and receiving in themselves that recompence of their error which was meet.

28 And even as they did not like to retain God in their knowledge, God gave them over to a reprobate mind, to do those things which are not convenient;

29 Being filled with all unrighteousness, fornication, wickedness, covetousness, maliciousness; full of envy, murder, debate, deceit, malignity; whisperers,

30 Backbiters, haters of God, despiteful, proud, boasters, inventors of evil things, disobedient to parents,

31 Without understanding, covenant breakers, without natural affection, implacable, unmerciful:

32 Who knowing the judgment of God, that they which commit such things are worthy of death, not only do the same, but have pleasure in them that do them."

Let's Get Serious

We believe it is only fair to remind you that the two-fold revelation we've been writing about is given to all men, including savages who run around naked in rain forests and who have never seen a Bible, heard a Gospel preacher or ever heard the name of God.

All of us have had the benefit of 2,000 years of Christian history. We have Bibles, we've heard of Jesus Christ, and we drive by a dozen churches on our way to work every day.

We have Christian radio broadcast, Christian television programs, and Christians who, like me, have gone out of our way to inform you of God and Jesus Christ.

There is no man or woman on earth who has an excuse not to know God and not believe in Jesus Christ.

If we respond positively to God's two-fold revelation, He will give us more spiritual light. If we reject God's two-fold revelation, God will let us have our own way and we will live the rest of our lives in spiritual darkness and die without God and without hope.

Richard Dawkins Has a Closed Mind

Much more recently, Richard Dawkins, during a debate with Cardinal George Pell, was asked by the moderator what kind of proof would change his mind about God's existence. Dawkins indicated in his response that:

"Even if a great big giant 900-foot-high Jesus . . . strode in and said, 'I exist. Here I am,' is mind likely still would not be changed." [10]

---

[10] Ham, Ken (2012-10-01). The Lie: Evolution (Revised & Expanded) (Kindle Locations 503-506). Master Books. Kindle Edition.

An Invitation

Would you like to become a true follower of Jesus Christ?

If you are willing to repent of your sins and believe the Gospel, you can be saved right now, wherever you are while you are reading this.

To repent of your sins means that you are sorry for your sins and you are willing, with God's help, to turn from them.

Would you like to have your sins forgiven and be saved?

If so, pray the following prayer from your heart.

Why not do it right now? The following is a suggested prayer.

Dear God,

I know that I am sinner.

I believe that you died for me.

I believe you arose from the grave and are alive right now.

Right now, Jesus, I want to turn from my sins and be saved.

Please forgive me. Come into my heart and save me.

I will do my best to live for you from now until the day I die.

In Jesus' Name, Amen.

Chapter 12

It's Time to Decide Where You Stand

You are either on the front lines in the battle for the minds and souls of your children, your church, and your country, or you and your children have already been taken prisoners of war by the lost world.

The debate is not over *Science vs the Bible.* The real debate is between *Science and Evolution.* We don't need to bring religion or the Bible into it. This is a debate between *True Science and Junk Science.*

The Theory of Evolution is Junk Science and people need to know it. The problem is, evolution is the only concept students in public schools are allowed to hear.

It is a closed system and neither students nor their teachers have been allowed to even be exposed to anything but that. That's why there's so much ignorance on the subject.

1. Charles Darwin was not a Scientist.

Darwin was the son of a doctor. He dropped out of Medical school after two years and changed majors and earned a degree in Divinity.

We call Darwin an apostate Divinity student. Yet, he is the author of the Origin of the Species, which is the Bible of the Evolutionists.

2. Evolution is Based on Blind Faith - Not Science.

Evolution is not happening today, nor was it ever observed by anyone, at any time, in the recorded history of man. It is a theory based on blind faith. It is widely accepted today because it is the only explanation students are allowed by law to hear. True science is based observation and experiment.

3. Evolutionist Can't Tell Us When, Where, or How Life Began.

Their theories require too much faith for this present writer to accept.

4. The Magic formula: Nothing x Time x Chance = Everything is Not Scientific. It is Wishful Thinking.

5. Spontaneous Generation is Not Scientific. It is accepted by Blind Faith.

6. The Fossil Record Does Not Support Evolution.

The fossil record reveals that life appeared suddenly and when it appeared it was already diversified and complex. There are no transitional forms in the fossil record.

7. There is not one Scientifically Observed and Recorded case of Transmutation of Species in the History of Science.

8. There are Hundreds of Reputable Scientists who Do Not Believe in the Theory of Evolution.

Sir Author Keith

"Evolution is unproved and unprovable. We believe it because the only alternative is special creation and that is unthinkable. [11]

Professor D.M.S. Watson

"Evolution is accepted by zoologists, not because it was ever observed to occur, or can be proved by logically coherent evidence to be true, but because the only alternative, special creation, is clearly incredible." [12]

---

[11] W.A. Criswell, "did man just Happen, published by not edited by the School of tomorrow. P.O. Box 29000, Lewisville, Texas. Page 54

[12] Criswell, Did Man Just Happen, Page 54

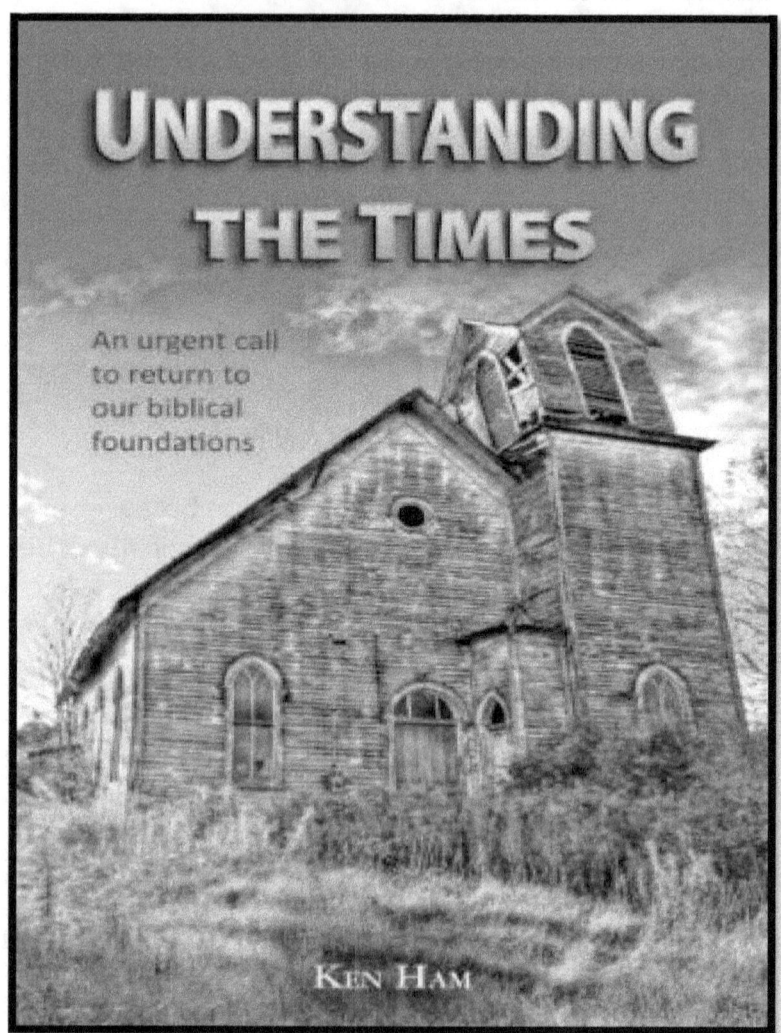

ANSWERING
THE
NEW ATHEISM

DISMANTLING
DAWKINS' CASE
AGAINST GOD

SCOTT HAHN · BENJAMIN WIKER

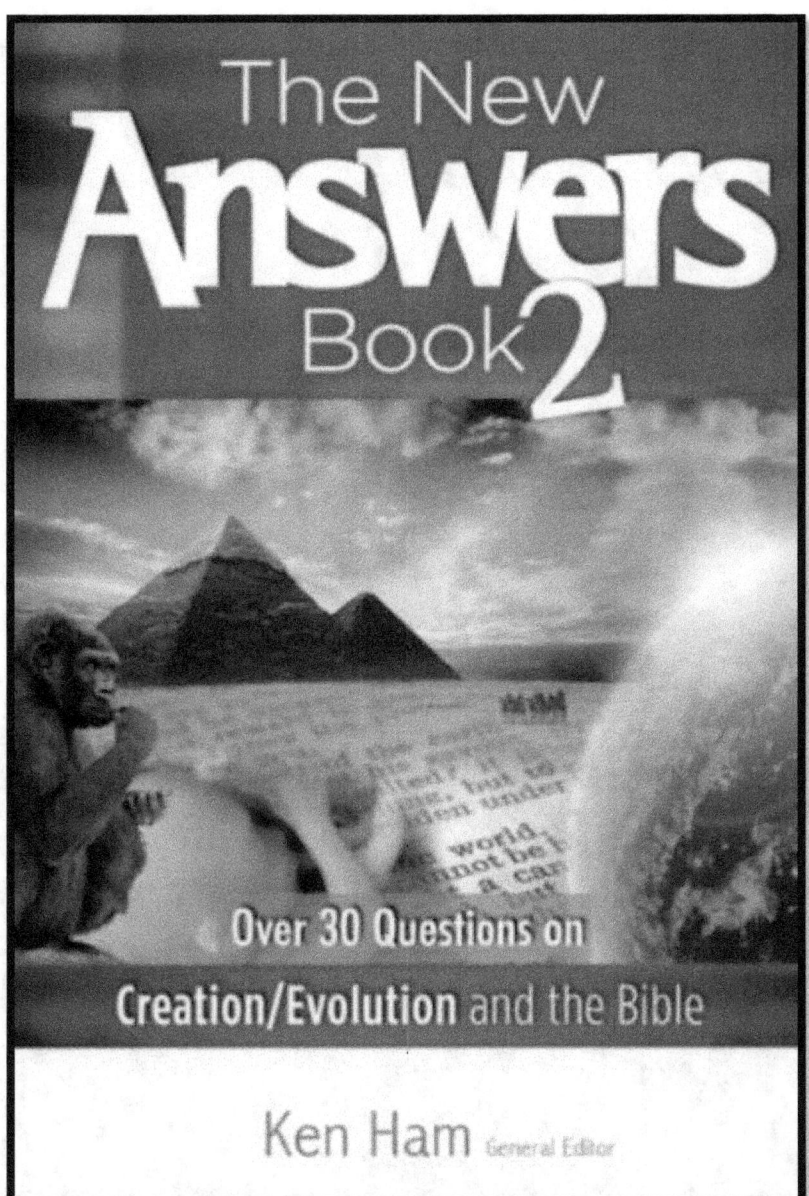

The New

Answers

Book 2

Over 30 Questions on

Creation/Evolution and the Bible

Ken Ham General Editor

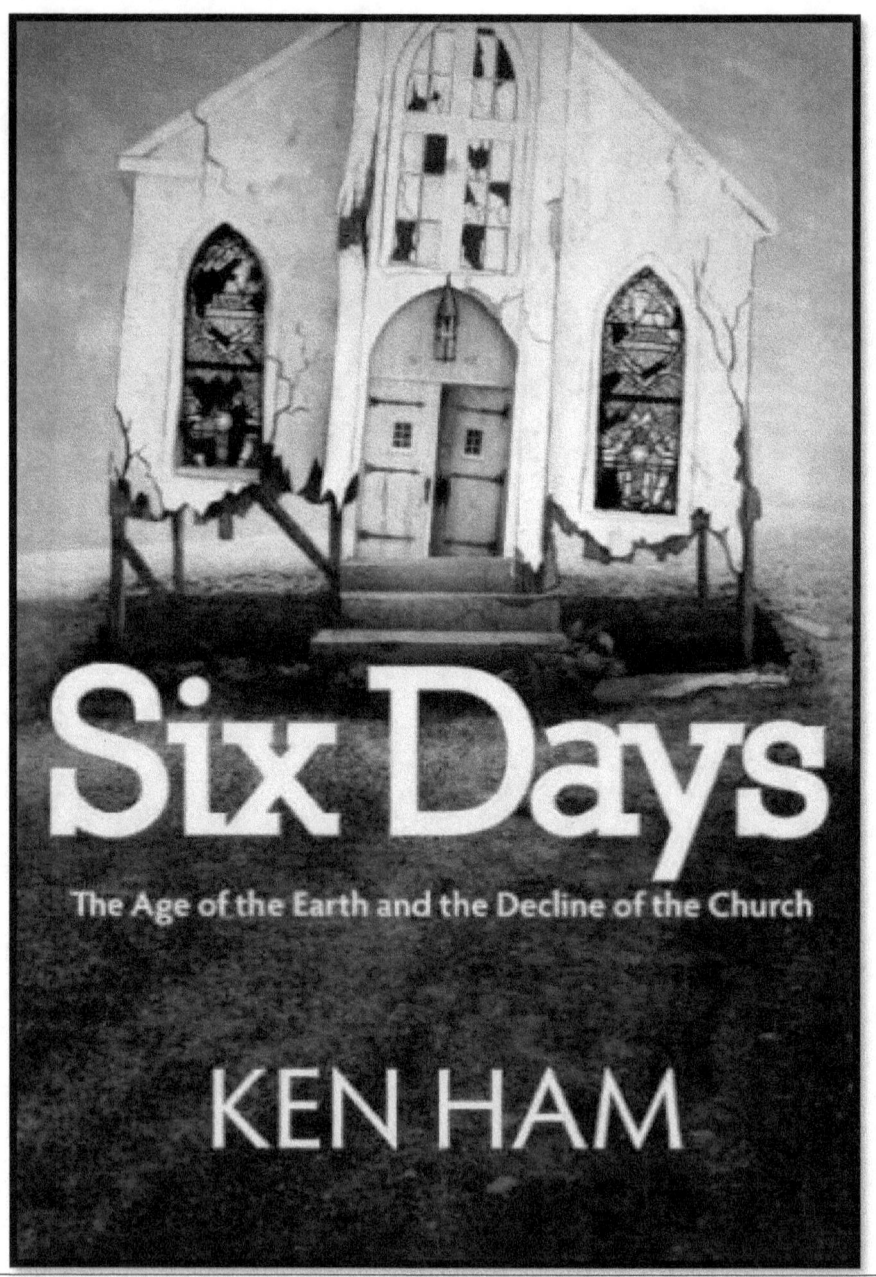

Six Days

The Age of the Earth and the Decline of the Church

KEN HAM

Raising Godly
Children in
an Ungodly
World

Leaving a
Lasting Legacy

Ken Ham   Steve Ham

Colin Gunn & Joaquin Fernandez Present

INDOCTRINATION

PUBLIC SCHOOLS AND THE DECLINE OF CHRISTIANITY

EDITED BY
CHARLES LAVERDIERE

CONTRIBUTORS INCLUDE:

R.C. Sproul, Jr.
Ken Ham
Doug Phillips
Voddie Baucham, Jr.
E. Ray Moore Jr.
Kevin Swanson
Israel Wayne
John Taylor Gatto
Samuel Blumenfeld
Erwin Lutzer

already gone

Why your kids will quit church — and what you can do to stop it

KEN HAM & BRITT BEEMER
WITH TODD HILLARD

THE LIE

THE LIE: Evolution/Millions of Years KEN HAM

The Great Dinosaur Mystery

A BIBLICAL VIEW OF THESE AMAZING CREATURES

SOLVED!

K e n   H a m

The Remarkable
RECORD of
JOB

*The Ancient
Wisdom, Scientific
Accuracy, & Life-Changing
Message of an Amazing Book*

HENRY M. MORRIS

How Do You Kill

11 MILLION PEOPLE?

WHY THE TRUTH MATTERS
MORE THAN YOU THINK

ANDY ANDREWS

NEW YORK TIMES BEST-SELLING AUTHOR of
THE NOTICER and THE TRAVELER'S GIFT

The Answers Book for **Kids**

Book for **Kids**

Volume 1

22 Questions from Kids on Creation and the Fall

DID MAN
JUST HAPPEN?

a pointed
answer for
evolutionists

by W. A. CRISWELL
introduction by PAIGE PATTERSON

About the Author

Gene Keith was born William Eugene Keith in Tarpon Springs, Florida, on December 25, 1932. His parents were Walter Keith Jr. Louise (Campbell) Keith. He graduated from Tarpon Springs High School in 1950, and married his sweetheart, Tuelah Evelyn Riviere in 1952. Tuelah's parents were Lawrence and Viola Riviere. Gene became Christian in 1952 and entered the ministry in 1953.

Florida Baptist Witness

The following material is from "Circling the Wagons," which appeared in the Florida Baptist Witness January 28, 2013 Joni B. Hannigan, Managing Editor).

Gene Keith is part of a ten-generation legacy of pioneer Christian leaders from Kentucky, Texas, and Florida. Since 1773 when John Keith hosted the first meeting of Virginia's Ten Mile Baptist Church, the Keith men for at least ten generations have led their congregations as Baptist preachers, elders or deacons, to be pioneers in sharing the Gospel. By wagon, on horseback, on foot, and by car, they've traveled carrying the Good News of Christ from the thick forests of Virginia, across the green mountains of Kentucky, to the High Plains of Texas before finally turning back southeast to settle in sun-drenched Central Florida where three generations now pastor two churches just 20 miles apart. "

Media

Gene has many years of experience in broadcasting, not only as a "DJ," but including his own daily talk show type program known as "The Sound of Inspiration," which was popular on several radio stations in North Central Florida. He had an afternoon show on the first FM Country Music station in Gainesville.

Education

Gene attended Stetson University, the University of Florida, and received his BA from Luther Rice College. He has many years of experience in the Christian School movement. He is the founder of the Countryside Christian School which celebrated its 45thAnniversary in 2019.

He has experience as a pastor and a Christian School Principal. Gene also served as a Consultant Field Representative for Accelerated Christian Education, during which time he helped establish several Christian Schools in Florida. Many of his family are involved in education serving as Principals and teachers.

Retired Pastor

Gene has served as pastor in Taft, Otter Creek, and the Pastor of the First Baptist Church of Cape Canaveral, Florida in 1968-1969, when America sent the first Astronauts to the moon. He served as the pastor of the Southside Baptist Church of Gainesville twice. Gene retired after 50 plus years in the ministry and is presently Pastor Emeritus of the Countryside Baptist Church of Gainesville, Florida. Gene turned 86 on Christmas Day 2018 and spends most of his time writing and speaking.

Published Author

Gene has more than twenty books already available and several in the process. His books are available from Amazon, both in Kindle format and in print.

Other Books by Gene Keith

You Can Understand the Revelation.

Daniel: The Key to Prophecy

Cremation: Are You Sure?

It's All About Jesus

Religious but Lost

Suicide: Is Suicide the Unpardonable Sin?

Getting Started Right: A Handbook for Serious Christians

Easter: Facts versus Fiction

How to Enjoy Christmas in a World that has lost its Way

Evolution: Facts versus Fiction

Why do Bad Things Happen to God's People?

The Radical Same Sex Revolution

Can a Saved Person Ever be Lost Again?

Otter Creek: True Stories of People and Places

Public School or Christian School?

One Nation Under God - or Allah. Can America Survive?

Financial Solutions (A Handbook for Church Leaders)

Our Story: God is Good- - All the Time!

Stop Changing History

The Pope, Peter, Walls, Guns, and Donald Trump

If you would like to order any of Gene's books, go to Amazon.com, type in Gene Keith Books, or the title of the book you desire and simply follow the links.

You may also correspond with Gene by email.

gk122532@gmail.com